THIS BOOK BELONGS TO:

_____

THE WONDERFUL WORLD OF SERVALS

MIMI JONES

Dedicated to all who love servals!

ISBN 978-1-958985-68-7

www.joeysavestheday.com

A Mimi Book

Servals are wild cats found in Africa.

They live in savannas, wetlands, and tall grasslands.

Servals are solitary animals and usually live alone.

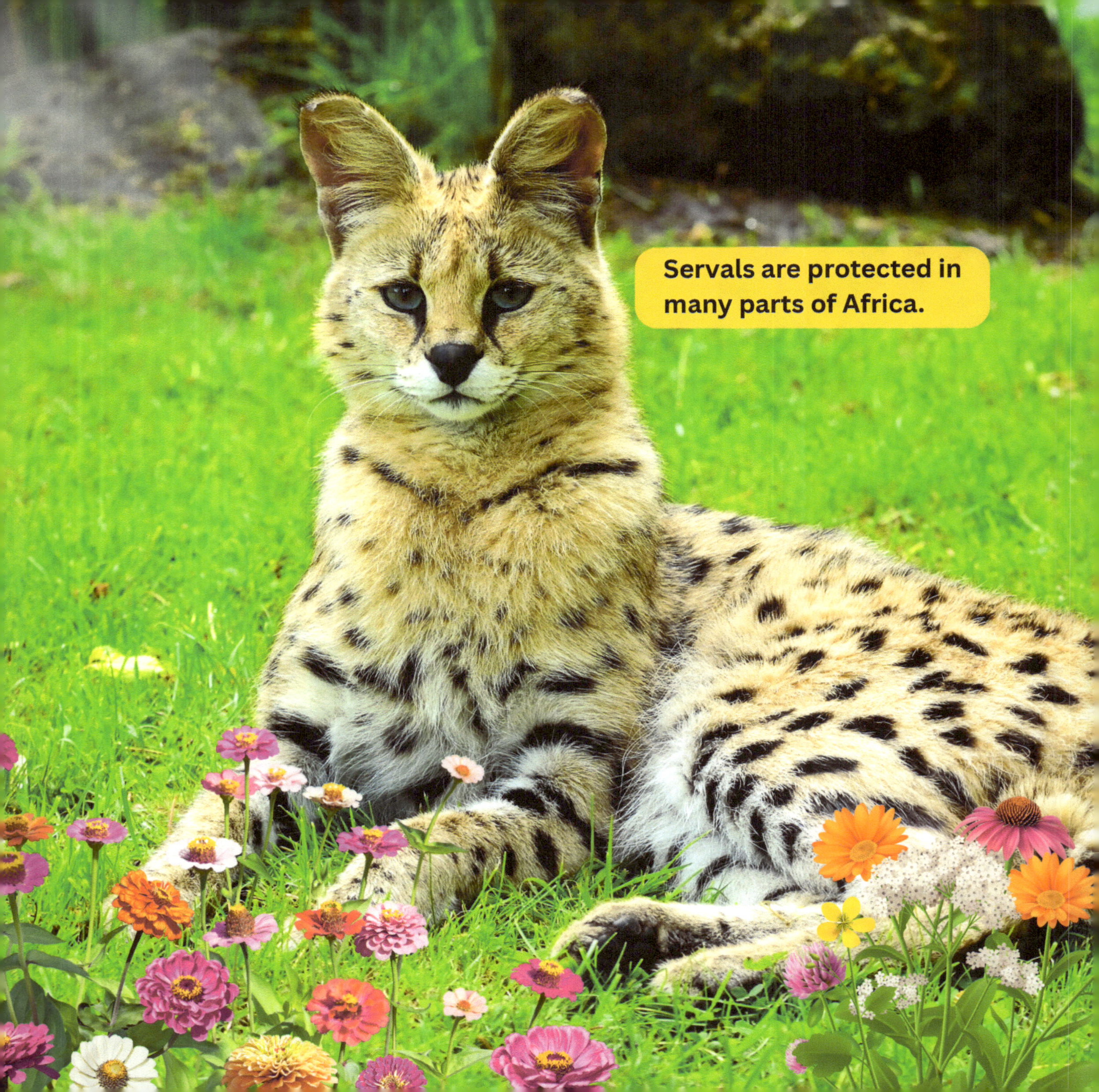

Servals are protected in many parts of Africa.

Servals have the longest legs compared to their body of any wild cat!

Their golden-yellow fur is decorated with bold black spots and stripes.

Their coat helps them blend into grass and shadows.

Their long necks help them see over tall grasses.

Servals have large ears that can rotate independently.

Their ears help them hear tiny sounds, like mice rustling in the grass.

Servals are amazing jumpers and can leap over 10 feet high!

10

They are fast runners and can reach speeds up to 50 miles per hour.

50

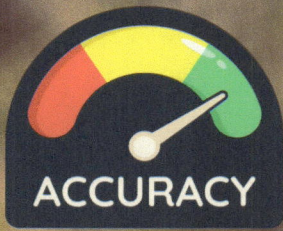

ACCURACY

Servals use a "stalk and pounce" method to catch prey. Their strong legs help them pounce with incredible accuracy.

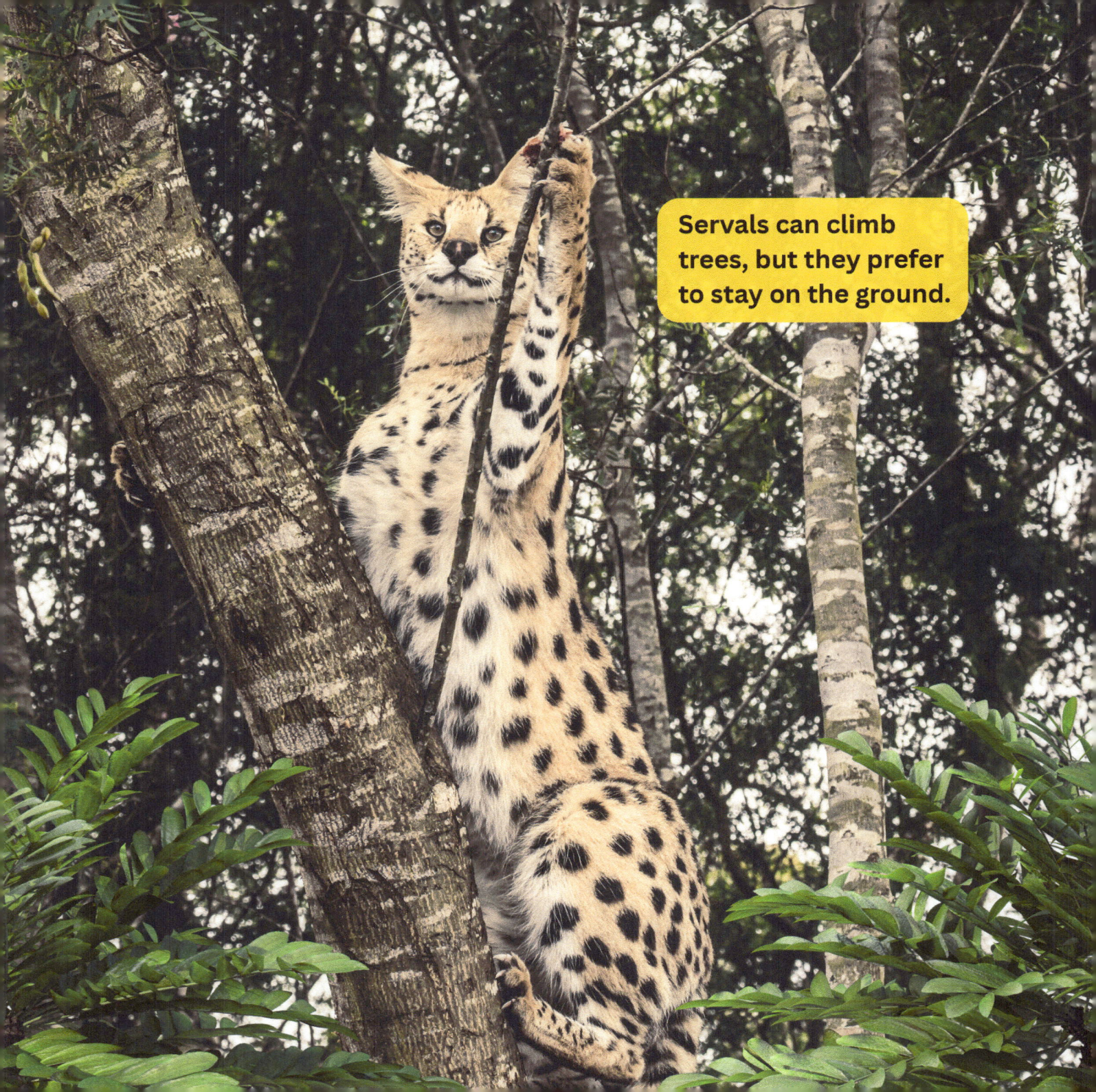

Servals can climb trees, but they prefer to stay on the ground.

SWIM

They are great swimmers but usually avoid water unless necessary.

Servals are nocturnal, meaning they are mostly active during the night.

They are carnivorous and eat small animals such as birds, frogs, insects, and rodents. A single serval can eat up to 4,000 rodents in one year!

4000

1 YEAR

Baby servals are called kittens.

Female servals have litters of one to four kittens.

1 – 4

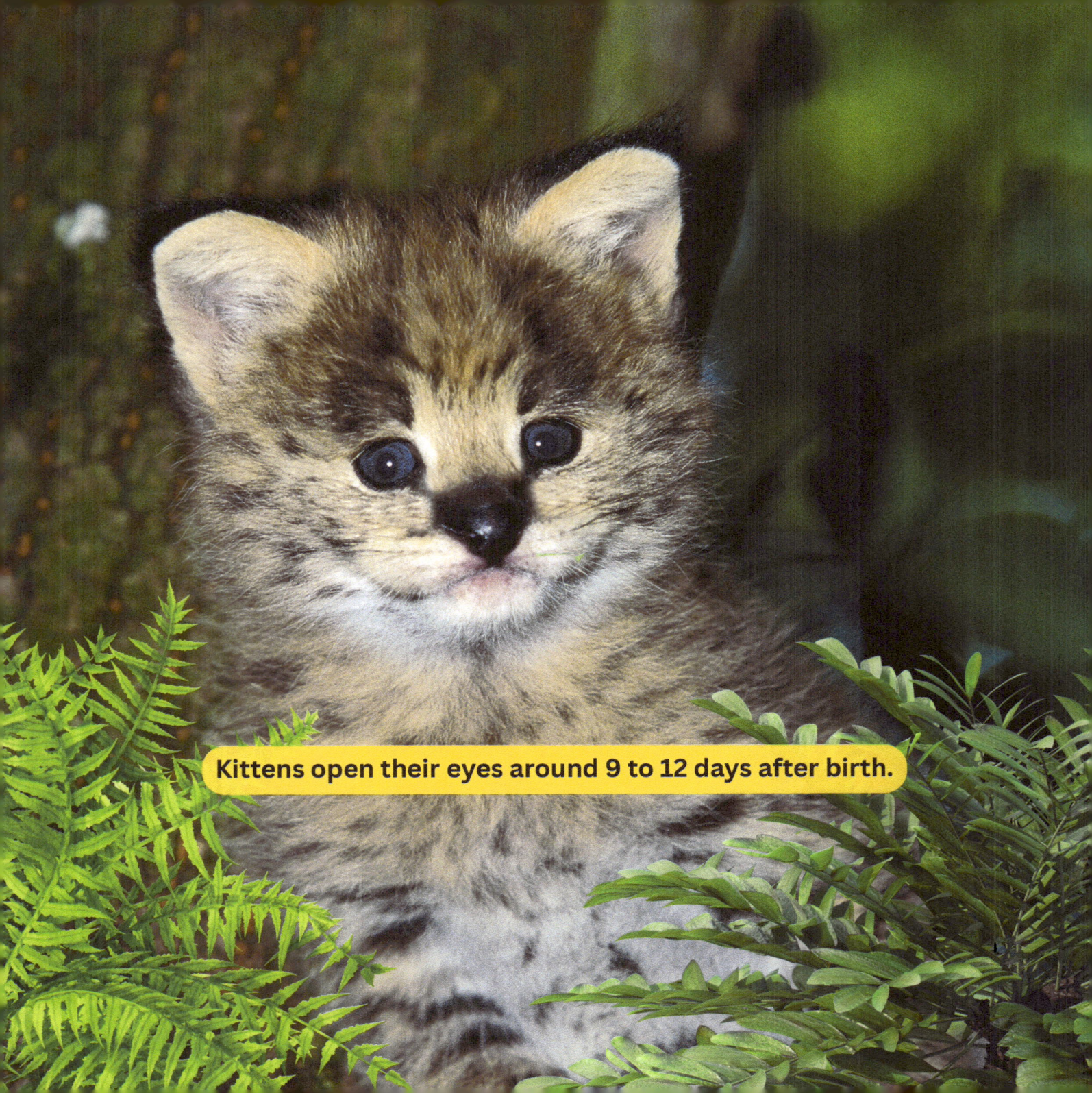

Kittens open their eyes around 9 to 12 days after birth.

6
MONTHS

Serval moms hide their babies in tall grass or empty burrows. Serval kittens start hunting on their own around six months old.

GROWL

CHIRP

MEOW

HISS

Servals chirp, hiss, growl, and meow, but they don't roar. They use tail flicks and ear movements to communicate.

Servals clean themselves just like house cats do by licking their fur.

Servals are distant cousins of both lions and cheetahs.

Servals have been important to African ecosystems for thousands of years.

Their population is stable, but habitat loss can affect them.

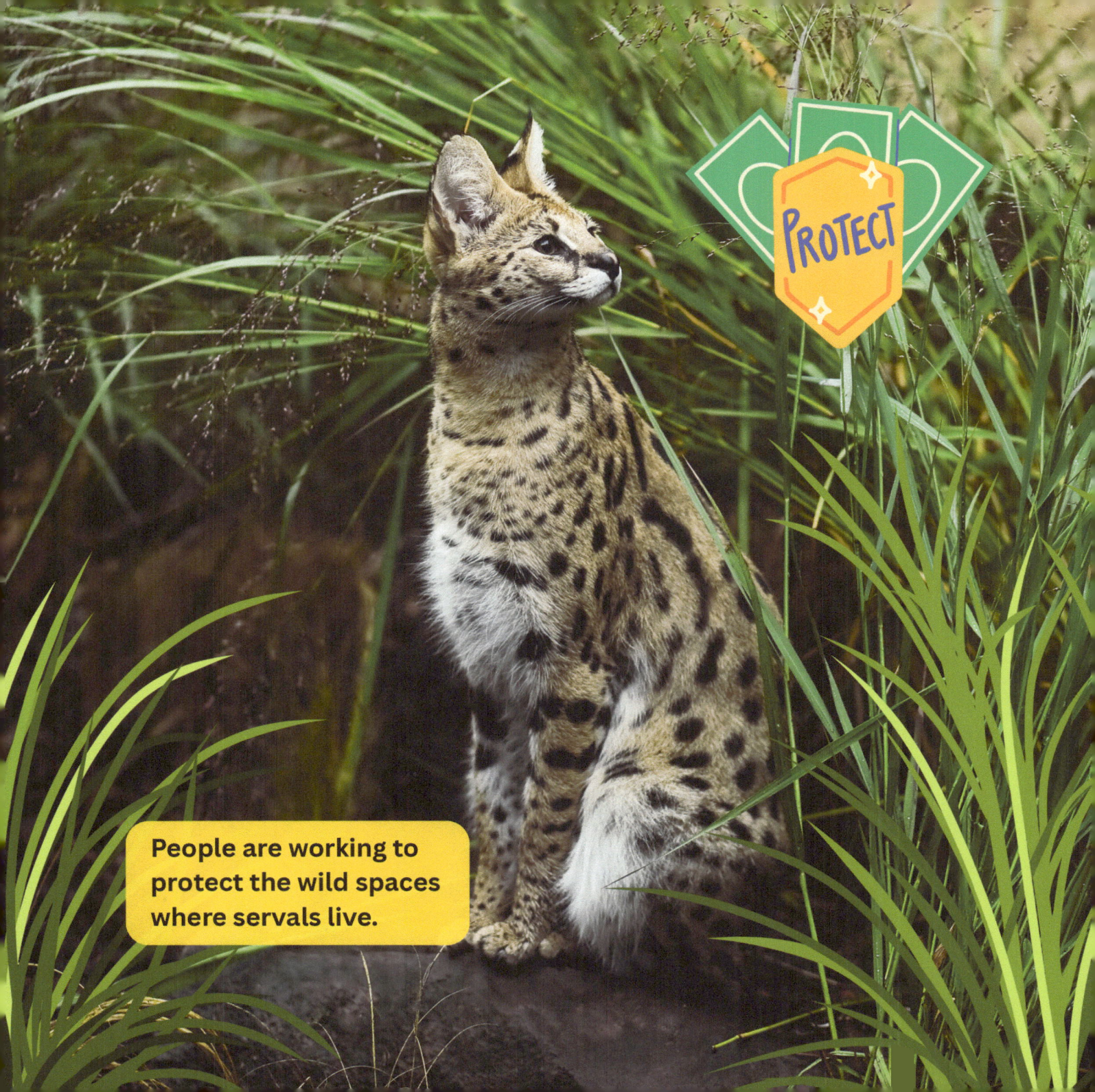

PROTECT

People are working to protect the wild spaces where servals live.

Count the servals!

Thanks for exploring the wild and wonderful world of servals with us! We hope you had as much fun as a serval spotting a sneaky mouse.

If you enjoyed this book, tell your friends and maybe leave a little review.

www.ingramcontent.com/pod-product-compliance
Lightning Source LLC
Chambersburg PA
CBHW060839270326
41933CB00002B/132